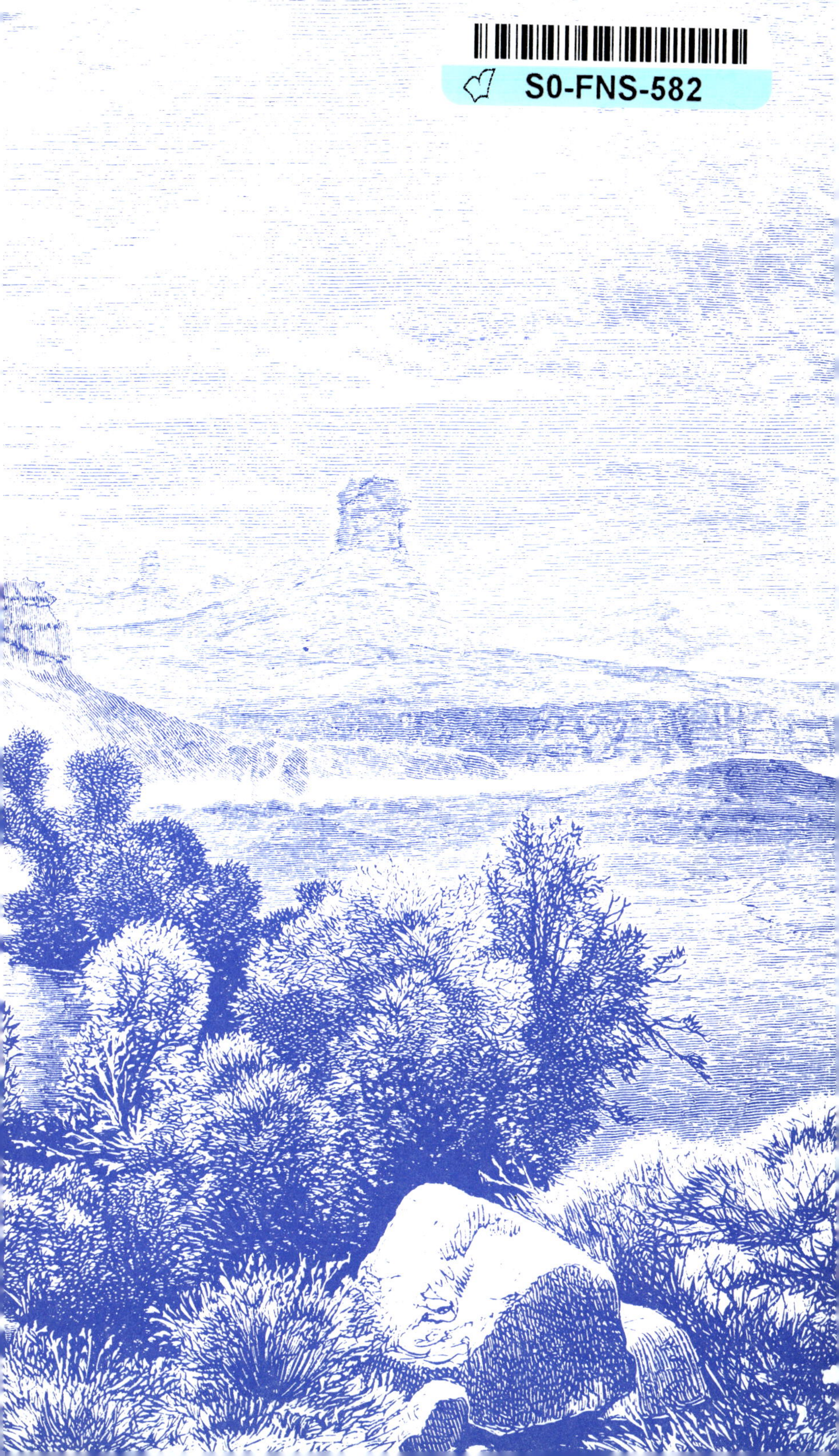

A GOLD HUNTER'S EXPERIENCE

A GOLD HUNTER'S EXPERIENCE

BY

CHALKLEY J. HAMBLETON

CHICAGO
PRINTED FOR PRIVATE CIRCULATION
1898

A GOLD HUNTER'S EXPERIENCE

BY

CHALKLEY J. HAMBLETON

YE GALLEON PRESS
FAIRFIELD, WASHINGTON

Library of Congress Cataloging-in-Publication Data

Hambleton, Chalkley, J., b. 1829.
　A gold hunter's experience.

　Reprint. Originally published: Chicago : Printed by R.R. Donneley, 1898.
　Includes index.
　I. Hambleton, Chalkley J., b. 1829. 2. Pikes Peak Region (Colo.)—Gold discoveries. 3. Gold mines and mining—Colorado—Pikes Peak Region— History—19th century. 4. Pioneers—Colorado—Pikes Peak Region—Biography. I. Title.

F782.P63H36　　　　1988　　　　978.8'56　　　　88-17086
ISBN 0-87770-454-6

I have often been asked to write an account of my Pike's Peak Expedition in search of gold. The following attempt has been made up partly from memory and partly from old letters written at the time to my sister in the east.

　　　　　　　　　　　　　　　　　　　　　C. J. H.

HOUSTON PUBLIC LIBRARY

R0121205665
SSCCA

A GOLD HUNTER'S EXPERIENCE

A Gold Hunter's Experience

Early in the summer of 1860 I had a bad attack of gold fever. In Chicago the conditions for such a malady were all favorable. Since the panic of 1857 there had been three years of general depression, money was scarce, there was little activity in business, the outlook was discouraging, and I, like hundreds of others, felt blue.

Gold had been discovered in the fall of 1858 in the vicinity of Pike's Peak, by a party of Georgian prospectors, and for several years afterward the whole gold region for seventy miles to the north was called "Pike's Peak." Others in the East heard of the gold discoveries and went West the next spring; so that during the summer of 1859 a great deal of prospecting was done in the mountains as far north as Denver and Boulder Creek.

Those who returned in the autumn of that year, having perhaps claims and mines to sell, told large stories of their rich finds, which grew larger as they were repeated, amplified and circulated by those who dealt in mining outfits and mills. Then these accounts were fed out to the public daily in an appetizing way by the newspapers. The result was that by the next spring the epidemic became as prevalent in Chicago as cholera was a few years later.

Four of the fever stricken ones, Enos Ayres, T. R. Stubbs, John Sollitt and myself, formed a partnership, raised about $9,000 and went to work to purchase the necessary outfit for gold mining. Mr. Ayres furnished a larger share of the capital than any of the others and was not to go with the expedition, but might join us the following year. Mr. Stubbs and I were both to go, while Mr. Sollitt was to be represented by a substitute, a relative whose name was also John Sollitt, and who had been a farmer and butcher and was supposed to know all about oxen. Mr. Stubbs was a good mechanic, an intelligent, well-read man, and ten years before had been to California in search of gold.

Our outfit consisted of a 12-stamp quartz mill with engine and boiler, and all the equipments understood to be necessary for extracting gold from the rock, including mining tools, powder, quicksilver, copper plate and chemicals; also a supply of provisions for a year. The staple articles of the latter were flour, beans, salt pork, coffee and sugar. Then we had rice, cornmeal, dried fruit, tea, bacon and a barrel of syrup; besides a good supply of hardtack, crackers and cheese for use while crossing the plains, when a fire for cooking might not be found practicable. These things were all purchased in Chicago, together with the fourteen wagons necessary to carry them across the plains. Then all were shipped by rail to St. Joseph, Mo., where the oxen were to be purchased. The entire outfit when loaded on the cars, weighed twenty-four tons.

I stayed in Chicago till the last to help purchase and forward the outfit and supplies, while Stubbs and Sollitt (the substitute) went to St. Joe to receive and load them on the wagons and to purchase the oxen.

On the 1st day of August, all was ready, and we ferried our loaded wagons and teams across the Missouri River into Kansas to make a final start next morning into regions to us unknown. Stubbs started the same day by stage for the mountains, to prospect and look out for a favorable location and then to meet the train when it arrived at Denver. Sollitt was to be trainmaster, which involved the oversight and direction of the teams and drivers, and the duty of frequently going ahead to pick out the best road and select a favorable place to camp at night, where water and grass could be had. I was the general business man of the expedition, had full power of attorney from Mr. Ayres to represent and manage his interest, and hence I had the control and responsibility in my hands and practically decided all important questions relating to the business.

The fourteen ox-drivers were all volunteers, who drove without pay—except their board—for the sake of getting to the gold regions to make their fortunes there. Most of them were from Chicago—three married men who left families behind, and one a young dentist. Another was the son of a prominent public woman who was a rigid Presbyterian, and when I left Chicago his father gave me a satchel full of religious books to give to him in St. Joe to read on the plains. He deliberately pitched them into a loft, where they were left. Another was a young Illinois farmer, named Tobias, a splendid fellow. Among those we secured in St. Joe were one German and two Missourians.

The principal article in the outfit of each individual, aside from his ornaments in the shape of knives and pistols, was a pair of heavy blankets. One of the Missourians first appeared without any, but next morning he had a quilted calico bed cover, stuffed with cotton, borrowed probably from a friendly clothesline, and which, at the end of the journey, presented a very dilapidated appearance.

Early in the morning of August 2d

all were busy yoking oxen and hitching them to the wagons, but as most of the drivers were green at the business and did not know "haw" from "gee," and a number of the oxen were young and not well broken, it was several hours before our train was in motion and finally headed for "Pike's Peak." The train consisted of fourteen wagons, a driver for each, forty yoke of oxen, one yoke of cows and one pony with a Mexican saddle and a rawhide lariat thirty feet long, with an iron pin at the end to stick in the ground to secure the animal.

For the first two or three miles, while crossing the level valley, all went well, but when we reached the bluffs and ravines that bounded the river valley on the west, the green oxen began to balk and back and refused to pull their loads up the hills, and the new drivers were nonplused and helpless. The better teams went ahead and were soon out of sight, while the poorer ones had to double up, taking one wagon up a hill and then going back for another, and consequently made slow progress. Instead of riding or walking along like a "boss" at ease, I soon found myself fully occupied in whipping up the poorly broken oxen on the off side, while the green drivers whipped and yelled at those on their side of the team. It was surprising how soon the nice city boys picked up the strong language in use by teamsters on the Western plains. The teams got separated, and the train stretched out two or three miles long. Then Sollitt rode ahead, picked out a camping place, and directed the drivers to halt and unyoke as they reached it; but when it became dark three or four teams were still from a quarter of a mile to a mile behind, and in trouble, so they unhitched the oxen and let them run in their yokes for the night.

A GOLD HUNTER'S EXPERIENCE

Our lunch and our supper that day consisted of crackers and cheese, as we had no time to cook.

About dark a shower came up, and it drizzled a good part of the night—the last rain we met with for many weeks. We rolled ourselves up in our blankets on the ground, under the wagons or in a small tent we had, for sleep. At daylight next morning we all started in different directions through the wet bushes that filled the ravines to find the scattered oxen, and before noon they were all collected at camp. We had hot coffee and some cooked things for breakfast. But several accidents had occurred. The cows had fallen into a gully with their yoke on and broken their necks, one load of heavy machinery had run down hill and upset, one axle, two wagon tongues, one yoke and some chains were broken. Sollitt, with two or three of the drivers who were mechanics, went to work to repair damages. As we seemed short of oxen, I rode back to St. Joe and bought two yoke more, spending the last of our money except about fifty dollars.

By next morning we were ready for a new start. Experience had already taught us something, and we adopted more system and some rules. All the teams were to keep near together, so as not to leave the weaker ones behind in the lurch. Our cattle were to be strictly watched all night by two men on guard at a time—not together, but on opposite sides of the herd. Two would watch half the night and then be relieved by two others who stood guard till morning. We all took our turns except the cook, who was relieved from that duty and from yoking and hitching up his own team, as cooking for sixteen men while in camp was no sinecure. The man chosen for cook was one of the drivers from Chicago named Taylor, who had cooked for campers and for parties at work in the woods. He was really a good plain cook. His utensils consisted of some large boiling pots and kettles, a tin bake oven, two or three frying pans, a two-gallon coffeepot and a few other usual articles.

Each person had a tin plate, a pint tin cup with a handle, and an iron knife, fork and spoon. The food was placed in the dishes and cups on the ground, and while eating we stood up, sat on the ground or reclined in the fashion of the ancient Romans, according to our individual tastes. The article of first importance at a meal was strong coffee and plenty of it. Next came boiled beans with pork, whenever there was time to cook them; and that could generally be done during the night. Then we had some kind of bread, cake or crackers, and sometimes stewed dried fruit.

About the third day out our open air

prairie appetites came, and it seemed as if we could eat and digest anything. I had been a little out of health for some time, was somewhat dyspeptic, and had not tasted pork for years. Soon I could devour it in a manner that would have shocked my vegetarian friends; and for the next two years I was conscious of a stomach only when hungry.

The third day the teams went a little better, but we had to double up sometimes to pull the wagons up the hills and out of the deep gullies we had frequently to cross, so we only made seven or eight miles. In a few days we got out on the level prairie and went along faster. But every morning for a week, one or more of our cattle would be lost from the herd. They would sneak away during the night and hide in the bushes and ravines, or start back toward home. As I had no special duties in camp, or in yoking up in the

A GOLD HUNTER'S EXPERIENCE

morning, hunting them fell to my lot. If not found in the first search before starting time, I would ride back on the pony for miles, scour the country and hunt through the gullies and bushes for hours till the lost animal was found; then drive him along until the train was overtaken. That could easily be followed by the tracks of the wheels on the prairie. Hiawatha, Kansas, and a few scattered cabins some miles to the west of it were about the last signs of settlement and civilization that we saw.

That season was a very dry one in Kansas and on the Western plains. The prairies were parched and looked like a desert, except a fringe of green along the water courses. The heat was intense and the distant hills and everything visible seemed quivering from its effects. The dry ground and sand reflected the sun's rays into our faces, till a few with weak eyes were seriously affected. The iron about the wagons, and the chains were blistering to the touch. The southwest wind was like a blast from a heated furnace. It was worse than stillness, and I frequently took shelter behind a wagon to escape its effects.

This heat was very trying and debilitating to the oxen. They would pant, loll their tongues out of their mouths, refuse to pull, and lie down in their yokes. Sometimes we were compelled to keep quiet all day, and drive in the early evening and morning, and during the night when we could find the way. The most important thing was to find water near which to camp. Wolves began to surround our camp and the herd of oxen at night, and break the silence by their piercing howls. After we had gone to sleep, they would sneak into camp to pick up scraps left from supper, then come within a few feet of some one rolled up in his blanket and startle him with a howl. But with all their noise these prairie wolves were great cowards, and would run from any movement of a man.

Soon after starting out one evening for a night drive, after a very hot day, one of the weak oxen lay down and refused to go. That the train might not be delayed, they tied his mate to a wagon, and I concluded to stay behind with him till morning to see if he would recover. Soon after dark the wolves seeming to divine his condition and the good meal in store for them, collected around us a short distance off, and seated on their haunches, with howls of impatience waited for the feast. They were plainly visible by their glaring, fire-like eyes. I varied the monotony of the long night by walking around, sitting down, lying upon the ground, and occasionally falling asleep beside the sick ox. Then the wolves emboldened by the stillness, would sneak up close to us and break out in piercing howls, but they would instantly vanish when I got up and threw something at them.

Daylight came at last; the ox had grown worse instead of better, and I left him to his fate and the wolves, and followed the wagon tracks till I overtook the train in camp, early in the day, with an appetite for a quart of strong coffee and something to eat.

In this hot weather the oxen with their heavy loads did not make more than a mile an hour when on the march, so with the numerous delays it was nearly two weeks before we reached Marysville on the Big Blue River. This was a small settlement on the verge of civilization, with a few ranches, saloons and stores, situated on that branch of the old Oregon trail which started northward from Westport, Mo., and passed near Fort Leavenworth, Kan. The inhabitants had the reputation of being mostly outlaws, blacklegs and stock thieves. Their reputation inspired us with such respect for them that we kept extra watch over our cattle and possessions while in the vicinity.

About a week after starting, one of the drivers got homesick, discouraged and disgusted with the trip, left us and started back home on foot. This compelled Sollitt and me to drive his team. One of our wagons not being made of properly seasoned wood, became shaky from the effects of the heat and dry air of the plains. At Marysville I traded it off to a ranchman for a yoke of oxen and had the load distributed on the other wagons so that again we had as many drivers as teams. I also traded some of our younger, weaker oxen for old ones that served our purpose better, though they were of less market value.

We learned that between this place and the Little Blue, there was no water to be found to enable us to camp for a night, so we were compelled to make the trip—some twenty miles—at a single drive. As the weather was hot we started late in the afternoon, drove all night, and arrived early next day, at that small river, where we found water and grass. Sollitt rode ahead much of the time to pick out the road.

Our course for several days was now along the Little Blue in a northwest direction, toward Fort Kearney on the Platte. To avoid the side gullies and ravines, which were water courses in the spring, though now dried up, we frequently circled off two or three miles on to the level prairie, but had to return near the stream when we camped, in order to get water.

One day, off to the west, a mile or two away, we saw a single buffalo which had probably been outlawed and driven from the herd to wander in solitude over the plains. Our pony had

A GOLD HUNTER'S EXPERIENCE

crossed the plains before and was well used to buffalo. Sollitt mounted him, and, rifle in hand, rode for the lone beast. When approached he began to run, but the horse soon overtook him, and he received a bullet. Then he turned savagely on the horse and rider, and, with head down, chased them at high speed before trying to escape. The horse overtook him a second time and he received another bullet. Then he charged after the horse and rider again. When the horse's turn to chase came next, the buffalo received a third shot and soon fell dead. This was quite exciting sport for us "tenderfeet" who had never seen a buffalo hunt.

Sollitt, who was a butcher by trade, was now in his glory. He rode back to camp, sharpened his knives and with the help of one or two of the men carved up the animal and brought back a supply of fresh meat. This proved rather tough as the animal was an old bull, nevertheless the tongue and the tenderloin were relished, after having eaten only salt pork for three weeks.

The small stream of water in the Little Blue grew less and less as we approached its source, and the last night that we camped near it, there was no running water at all. The little that was to be seen stood in stagnant pools in the bottom of the river bed. When we would approach these pools, turtles, frogs and snakes in great variety, that had been sunning themselves on the banks, would tumble, jump and crawl into the water, and countless tadpoles wiggled in the mud, at the bottom, so that the water was soon black and thick. Its taste and smell were anything but appetizing. The oxen, though without water since morning, refused to drink it, even after we had dipped it up in pails and allowed it to settle. We boiled it for the coffee, but the odor and flavor of mud still remained. The situation had become serious and our only hope was to reach the Platte river before the oxen were famished from thirst. Earlier in the season, before the streams dried up, this was a favorite route of travel, but it was not so at this time of year and we saw very few passing teams.

By daylight next morning the oxen were yoked and hitched up and we commenced a forced march for water and salvation. The old trail seemed still to follow the course of the dried-up stream, bearing much to the west. We concluded to leave it and steer more to the north with the hope of striking the Platte at the nearest point. The prairie was hard and level, the day not excessively hot, and everything was favorable for a long drive. The rule for keeping together was ignored and each team was to be urged to its best speed, in the hope that the strong and the swift would reach the goal though the weak and the weary might fall by the way.

Before noon the teams were much separated. They halted for a nooning; the oxen browsed a little on sage brush and dried grass; the men lunched on crackers, cold coffee and the remnants of breakfast, but our water keg was empty. By the time the last team was at the nooning place, the head ones were ready to start on.

Sollitt rode ahead to explore and pick out the road, carrying his rifle on the saddle, as we were liable at any time to meet bands of treacherous, pillaging Pawnees, whose haunts were on the lower Platte. I formed the rear guard with the hindmost wagon, so that it would not be deserted and alone in case of accident. Each team was always in sight of the next one ahead of it, though the train was stretched out some three miles long. Late in the afternoon Sollitt rode back with the cheering news that he had seen the Stars and Stripes waving over Fort Kearney to the west and that he had picked out a camping ground near the river a few miles below. Soon after dark the last team was in camp and the men and beasts were luxuriating in the clear running water of the Platte.

The next forenoon we drove on to the fort and camped a mile or two west of it for a day's rest. This was on the 20th of August, so we had been out twenty days on the road from St. Joe. At the fort was a postoffice and here we received letters from our friends in the East, and spent a good part of the day in writing, in response to them. Letters were brought here by the coaches of the overland express which carried the United States mail to California.

The fort consisted of a few buildings surrounded by a high adobe wall for protection; and adjoining was a strong stockade for horses and oxen. There were a few United States troops here. Just outside the fort grounds were some ranches, stores, saloons and trading posts. The two Missourians proceeded forthwith to get dead drunk and it took them till next day to sober up. By way of apology they said the whisky tasted "so good" after being so long without it. We had no whisky on our train. It was one of the very few that crossed the plains in those days without that, so considered, essential article in frontier life.

Personally, through the entire period of my "Pike's Peak" experience, I adhered strictly to my custom of not tasting spirituous or malt liquors, nor using tobacco in any form.

We were now on the main central route of travel from the States to the mountains, Salt Lake, California and Oregon. We saw teams and trains daily going in both directions, and

Kearney was a favorite place for them to stop over a day and rest. Our course now lay along the south side of the Platte, clear to Denver; and with the prospect of level roads and plenty of grass and water, we looked forward hopefully to a pleasant trip the rest of the way. The valley of the Platte is a sandy plain, nearly level, extending westward for hundreds of miles from Kearney, bounded on the north and the south by low bluffs, some four or five miles apart. Back of these lie the more elevated, dry plains extending to great distances.

Winding through this valley is the Platte river, a half a mile or more wide, with water from an inch to two feet deep, running over a sandy bottom and filled with numberless islands of shifting sand. The banks were lined with willows and cottonwood bushes and bordered in many places by green, grassy meadows, but trees were a rarity and for some two hundred miles we did not see one larger than a good sized bush.

The day we camped near Kearney we began to see buffalo in small groups off a few miles to the south and west. When I awoke next morning, soon after daylight, I saw a lone one quietly eating grass about half a mile from camp. I got out a rifle and went toward him, stooping or going on my hands and knees through the wet grass, till within good rifle shot. I then stood up, took deliberate aim just behind the shoulder, and fired. He gave a quick jump, looked around and started toward me on the run with head down, in usual fashion, for a charge. My thought was that I had hit, but not hurt him. I dropped into the grass and made my way on hands and knees as fast as possible toward camp, a little agitated. Losing sight of me the animal soon stopped, stood still a few

A GOLD HUNTER'S EXPERIENCE

minutes and then suddenly dropped to the ground. He had been shot through the heart.

This was my first and last buffalo, as sneaking up to them and shooting them down did not seem much more like sport than shooting down oxen. I was neither a sufficiently expert rider nor hunter to chase and shoot them on horseback. The one I shot was carved by Sollitt and one of the men, and furnished us fresh meat for breakfast and several meals thereafter.

During the day we passed a ranch, occupied by a man and his son, twelve or fourteen years old. The boy had eight or ten buffalo calves in a pen, which he said he had caught himself and intended to sell to parties returning to their homes in the East. He had a well-trained little pony, which he would mount, with a rope in hand that had a noose at the end, and ride directly into the midst of a small drove of buffalo, and while they scattered and ran would slip his rope about the neck of a calf and lead it back to the ranch. The calf would side up to the pony and follow it along as if under the delusion that it was following its mother. The man traded in cattle by picking up estrays and buying, for a song, those that were footsore and sick, keeping them till in condition and then selling them to passing trains that were in need.

We now began to see buffalo quite plentifully off to the southwest, in small groups, and in droves of twenty or more. Sometimes hunters on horseback, who had camped near Kearney, were indulging in the excitement of the hunt, chasing and shooting, and in turn being chased by the enraged animals. That evening we camped on the verge of the great herd that extended some sixty or seventy miles to the westward, and blackened the bluffs to the south, and the great plains beyond as far as the eye could reach. This great herd was not a solid, continuous mass, but was divided up into innumerable smaller herds or droves consisting of from fifty to two hundred animals each. These kept together when grazing, marching or running, the bulls on the outside and the cows and calves in the center. Sometimes these small herds were separated from each other by a considerable space.

This great herd had probably started northward from the Arkansas in the spring and had now reached the Platte, where they lingered for water and the better grass that was found along the river. Following in the wake and prowling on the outskirts of this slowly moving host, were thousands of wolves, collected from the distant plains, to feast upon the young and the weakly, and the carcasses of those that were killed by accident or the hunter's gun.

A GOLD HUNTER'S EXPERIENCE

The turn for watching the cattle the first half of that night fell to the lot of two of the boys from Chicago. The cattle were grazing in a good meadow off toward the river, half a mile from camp. At dusk the boys went off to take charge of them. After dark the wolves began to howl in all directions and sometimes it sounded as if a hundred hungry ones were fighting over a single carcass. Then the buffalo bulls chimed in with the music and bellowed, apparently by thousands, at the same time. Pandemonium seemed to reign. The two boys got nervous, then frightened and finally panic-stricken, and long before midnight came rushing into camp declaring that they were surrounded by droves of hungry wolves and furious buffalo. The cattle were also disturbed and inclined to scatter and wander off.

Next morning early, all of us, except the cook, started off to hunt them up.

Some went up stream, some down, and some back along the road we had come. Tobias and myself waded the river to the north side to hunt them there, but we found neither cattle nor cattle tracks. We did find a huge rattlesnake, which we killed. The river was about three-quarters of a mile wide, and in no place over two feet deep. Wading it was easy enough if one kept moving, but if he stood still he would gradually sink into the quicksand till it was difficult to extricate his feet.

By noon, after this thorough search, we had collected all of our oxen but two, which could not be found. Sollitt was very suspicious of cattle thieves, and, whenever an ox was lost, his first opinion was that it had been stolen. Mine was that it had strayed off and hidden in some ravine or clump of bushes. He decided that these two lost ones had been taken by some ranchman or passing train. I believed they had gone off with the buffalo and that when they wanted drink badly they would come back to the river. I therefore concluded to let the train go on, while I, with the pony and some food, would stay behind and patrol the river for a day or two. I rode back eastward along the river's edge, searching in the bushes, and at night came to a ranch, near which I picketed the pony and slept on the ground. Next morning, after first examining the ranchman's cattle, I started westward again, making another thorough search as I went along. In the afternoon I found the stragglers quietly eating grass near the river, and then drove them along as fast as possible till the train was overtaken.

We were now right in the midst of the great herd, through which we journeyed for nearly five days. The anxiety they gave us was greater than that of any of our previous troubles. To avoid having the oxen stampeded, or run off with the buffalo at night, we wheeled our wagons into a circle when camping at the end of a day's drive, and thus formed a corral, into which we put as many oxen as it would hold, for the night, and chained the rest in their yokes to the wagon wheels on the outside. This was hard on the oxen, as they could not rest as well as when free, nor could they graze a part of the night, as was their habit. Whenever we looked off to the south or southwest, we would see dozens and dozens of the small droves of one or two hundred buffalo moving about in all directions. Some of the droves would be quietly eating grass, some marching in a slow, stately walk, and others on the run, going back and forth between their grazing grounds and the river. But each separate drove kept in quite a compact body.

Sometimes they would keep off from

the trail along which we traveled, for several hours at a time and not trouble us. At other times they would be going in such great numbers across our route, passing to and from the river, that we had to wait hours for them to get out of our way. Often a drove would get frightened at a passing wagon, the report of a gun, the barking of a dog, or some imaginary enemy, and would start on a run which soon became a furious stampede, the hindermost following those before them, and in their blind fury crowding them forward with such irresistible force that the leaders could not stop if they would. If they came suddenly to a deep gully the foremost would tumble in till it was full, and thus form a bridge of bone and flesh over which the rest would pass. Several times these frightened droves passed so near our wagons as to be alarming.

One drove came within a few yards of one of our wagons, and some of the drivers peppered them with bullets from their pistols. Though these frightened droves could not be stopped, they would shy to the right or left if an unusual commotion was made in time in front of them. When a drove, at some distance, seemed to be headed toward our train, we often ran toward it, yelling, firing guns, and waving articles of clothing. The leaders would shy off, and that would give direction to the whole body, and thus relieve us from danger for the time being.

Every teamster, traveler and hunter that crossed the plains felt that he must kill from one to a dozen or more buffalo. The result was that the plain was dotted and whitened with tens of thousands of their carcasses and skeletons. With this general slaughter and the increase of travel induced by the discovery of the Pike's Peak gold fields, no wonder that this was the very last year that these animals appeared in

A GOLD HUNTER'S EXPERIENCE

large numbers in the Platte valley. We always estimated their numbers by the million.* For some years after they appeared in large numbers in some parts of the great plains of the West, but they rapidly declined in number till they became extinct in their wild state.

While in their midst we not only had fresh meat at every meal, but we cut the flesh in strips and tied it to the wagons to dry and thus provided a small supply of "jerked" meat. In the dry, pure air of this region, though

---

*The estimate was probably not an exaggeration.
In a late work it is stated on the authority of railroad statistics that in the thirteen years from 1868 to 1881 " in Kansas alone there was paid out *two millions five hundred thousand dollars* for their bones gathered on the prairies to be utilized by the various carbon works of the country, principally in St. Louis. It required about one hundred carcases to make one ton of bones, the price paid averaging eight dollars a ton; so the above quoted enormous sum represented the skeletons of over thirty-one millions of buffalo."—*The Old Santa Fe Trail, by Col. Henry Inman p. 203.*

The author further says, "In the autumn of 1868 I rode with Generals Sheridan, Custer, Sully and others for three consecutive days through one continuous herd, which must have contained millions. In the spring of 1869 the train on the Kansas Pacific railroad was detained at a point between Forts Harker and Hays from nine o'clock in the morning until five in the afternoon in consequence of the passage of an immense herd of buffalo across the track."

Horace Greeley crossed the plains in 1859 in a stage coach, and as stated in his published letters, he saw a herd of buffalo that he estimated to contain over five millions.

in the heat of August, fresh meat did not spoil but simply dried up, if cut in moderate sized pieces. This was also found to be the case with fresh beef in the mountains. We felt relieved and heartily glad when the last drove of buffalo was left behind. Familiarity with them, as with the Indians, destroyed all the poetry and romance about them. They were not a thing of beauty. An old buffalo bull with broken horns and numerous scars from a hundred fights, with woolly head and shaggy mane, his last year's coat half shed and half hanging from his sides in ragged patches and strips flying in the breeze, the whole covered over with dirt and patches of dried mud, presented a picture that was supremely ugly.

On the journey from St. Joe to Kearney we found, along the water courses and ravines, enough of dry wood and dead trees to supply us plentifully with fuel for cooking and occasionally to light up the camp in the evening. To make sure of never being entirely out of wood, a small supply was carried along on the wagons. Along the Platte there was practically no wood to be had. For one hundred and fifty miles we did not see a single tree, but the buffalo supplied us with a good fuel called "buffalo chips," which was scattered over the plains in abundance, and which in this dry country, burned freely and made a very hot fire. When approaching camp in the evening, the drivers would pick up armsfull of fuel for the use of the cook and for the evening camp fire, and place it in a pile as they came to a halt.

As soon as we reached camp and while others were taking care of the oxen, the cook built a fire, drove two forked sticks into the ground, one on each side of the fire, placed a cross stick on them, and then hung his pots and kettle over the blaze. A big pot of beans with pork was boiled or warmed over. Coffee was prepared, and dough made of flour and baking powder was baked either in the tin oven or a Dutch oven. Frequently some of the men were seated on the ground around the fire, stick in hand with a piece of pork on the end of it, held near the coals to toast. While eating and during the early evening, talking, story telling and ironical remarks about the prolonged picnic — as the trip was called — were indulged in.

We were now on the main route of travel between the East and the Pike's Peak gold fields. Horse and mule teams going West, and traveling faster than our ox train could go, passed us frequently, and gave us the latest general news from the States. We also began to meet the vanguard of the returning army of disappointed gold seekers. They came on foot, on horse

A GOLD HUNTER'S EXPERIENCE

back and in wagons drawn by horses, mules and oxen, and many of them were a sorry, ragged looking lot. Judging from their requests from us, their most pressing wants were tobacco and whisky. In those days Western towns were full of enthusiastic, sanguine, roving men who were ever ready for any new enterprise, and they were the first to rush to the gold regions in the spring. But lacking pluck, perseverance and the staying qualities, they were the first to rush back when the difficulties and discouragements of the undertaking appeared in their way.

These returners told sad stories about life in the mountains, the prospects and the danger from Indians on the road. They said that there was but little gold to be found, that very few of the miners were making expenses, that food was scarce, and that before we reached our destination, nearly everybody there would be leaving for home. Besides, they said, there were hundreds of Indians along the route, robbing and murdering the whites. Such stories had a discouraging effect on some of our drivers and I was very fearful that a few of them would leave us and join the homeward procession.

Some of these chaps showed a humorous vein in the mottoes painted on the sides of their wagons. On one was "Pike's Peak or bust," evidently written on going out; under it was written, "Busted." On another was, "Ho for Pike's Peak;" under it was, "Ho for Sweet Home."

Each exaggerated account of the Indians made by these people, brought us nearer and nearer to them and made them seem more and more dangerous. Finally one morning as we reached the top of a gentle swell in the plain, a large band of them suddenly appeared in full view, camped at the side of

A GOLD HUNTER'S EXPERIENCE

our road about half a mile ahead of us. From all appearances there were five or six hundred or more of them. They belonged to the western branch of the Sioux tribe. We stopped a few minutes to consider the situation. We had heard and read enough about Western Indians to know that the safest thing to do was to appear bold and strong, while a show of weakness and timidity was often dangerous. So we placed in our belts all our ornaments in the shape of pistols and ugly looking knives, and those who had rifles carried them. Then we drove boldly forward toward the camp. I rode the pony beside the driver of the foremost wagon with my old shot gun in hand. Soon two or three of their mounted warriors or hunters rode at full speed toward us and then without stopping circled off on the plain and back to their camp. They were evidently making observations.

A GOLD HUNTER'S EXPERIENCE

Off to the north several hundred shaggy ponies were grazing in a green meadow near the river, and the greater part of their men seemed to be there with them. The camp was made up of some forty lodges, which looked like so many cones grouped on the plain.

These lodges were formed of poles, some fifteen feet long, the larger ends of which rested on the ground in a circle, while the smaller ends were fastened in a bunch at the top, with a covering of dressed buffalo skins stitched together. On one side was a low opening, which served for a door.

As we approached we were first greeted by a lot of dirty, hungry looking dogs, which barked at us, snarled and showed their teeth. Then there was a flock of shy, naked, staring children who at first kept at a safe distance, but came nearer as their timidity left them. The boys with their little bows and arrows were shooting at targets—

taking their first lessons as future warriors of the tribe.

When we got near the edge of the camp several of the old men came forward to greet us with extended hands, saying "how! how! how!" and we had to have a handshake all around. Some of them knew a few words of English. They asked for whisky, powder and tobacco. Instead, we gave some of them a little cold "grub." They looked over all the wagons and their contents, so far as they could, and were particularly interested in the locomotive boiler which was placed on the running gear of a wagon without the box, and with the help of a little rude imagination, somewhat resembled a huge cannon. I told them it was a "big shoot," and that seemed to inspire them with great respect for it. They looked under it and over it and into it with much interest.

The greater part of the squaws were

seated on the ground at the openings of their lodges, busily at work. Some were dressing skins by scraping and rubbing them, some making moccasins and leggings for their lazy lords, some stringing beads and others preparing food. The oldest ones, thin, haggard and bronzed, looked like witches. The young squaws, in their teens, round and plump, their faces bedaubed with red paint toned down with dirt, squatted on the ground and grinned with delight when gazed at by our crew of young men. We all traded something for moccasins and for the rest of the trip wore them instead of shoes.

Curious to see inside of the lodges, I took a cup of sugar and went into two or three under pretence of trading it for moccasins. Their belongings were lying around in piles, and the stench from the partly prepared skins and food was intolerable.

One old Indian seemed to think that

A GOLD HUNTER'S EXPERIENCE

I was hunting a wife, for he offered to trade me one of his young squaws for the pony. A pony was the usual price of a wife with these Western Indians. They exhibited no hostility whatever toward us. It might have been otherwise, had we been a weak party of two or three possessing something that they coveted.

They asked us if we saw any buffalo. When we told them that at a distance of two or three days' travel the plains were covered with them, they seemed greatly interested and before we got away began to take down some of their lodges and start off. They were out for their yearly buffalo hunt to supply themselves with meat for the winter. In moving they tied one end of their lodge poles in bunches to their ponies and let the other ends spread out and drag upon the ground, and on these dragging poles they piled their skins and other possessions. The young children and old squaws would often climb up on these and ride.

Cactus plants in hundreds of varieties grew in great abundance on these dry plains. They were beautiful to the eye, but a thorn in the flesh. As we walked through them their sharp needles would run through trousers and moccasins and penetrate legs and feet. We often ate the sickishly sweet little pears that were seen in profusion.

Prairie dogs by the million lived and burrowed in the ground over a vast region. The plains were dotted all over with the little mounds about two feet high that surrounded their holes. On these mounds the little animals would stand up and bark till one approached quite near, then dart into the holes. In places the ground was honeycombed with their small tunnels, endangering the legs of horses and oxen, which would break through the crust of ground into them. I shot at many of them, but never got a single animal, as they always dropped, either dead or alive, into the hole and disappeared from sight.

Many small owls sat with a wise look on top of these little mounds, and rattlesnakes, too, were often found there. When disturbed the owls and snakes would quickly fly and crawl into the holes. It was a saying that a prairie dog, an owl and a rattlesnake lived together in peace in the same hole. Whether the latter two were welcome guests of the little animal, or forced themselves upon his hospitality, in his cool retreat, I never knew.

One day we came to a wide stretch of loose dry sand, devoid of vegetation, over which we had to go. It looked like some ancient lake or river bottom. The white sand reflected the sun's rays and made it unpleasantly hot. The wheels sank into the sand and made it so hard a pull for the oxen that we had to double up teams, taking one wagon through and going back for another, so we only made about three miles that day.

The unexpected was always happening to delay us. The trip was dragging out longer than was first reckoned on, and the early enthusiasm was dying out. Walking slowly along nine or ten hours a day grew monotonous and tiresome. Then, after the day's work, to watch cattle one-half of every third night was a lonely, dreary task, and became intolerably wearisome. Standing or strolling alone, half a mile from camp, in the darkness, often not a sound to be heard except the howling of the wolves, and nothing visible but the sky above and the ground below, one felt as if his only friends and companions were his knife and his pistol.

In the early part of September violent thunderstorms came up every evening or night, with the appearance

A GOLD HUNTER'S EXPERIENCE

of an approaching deluge. Very little rain fell, however, but the lightning and thunder were the most terrific I ever saw or heard. There being no trees or other high objects around, we were as likely to be struck as any thing. For a few wet nights I crawled into one of the covered wagons to sleep, where some provisions had been taken out, and right on top of twelve kegs of powder. I sometimes mused over the probable results, in case lightning were to strike that wagon. We passed one grave of three men who had been killed by a single stroke of lightning. Graves of those who had given up the struggle of life on the way, were seen quite frequently along the route. They were often marked by inscriptions, made by the companions of the dead ones on pieces of board planted in the graves.

Now we came to extensive alkali plains, covered with soda, white as new fallen snow, glittering in the sunshine.

No vegetation grew and all was desolation. An occasional shower left little pools of water here and there, strongly impregnated with alkali, and from them the oxen would occasionally take a drink. From that cause, or some other unknown one, they began to die off rapidly, and within three days one-third of them were gone. The remainder were too few to pull the heavy train. The situation was such that it gave us great anxiety.

What was to be done? Either leave part behind and go on to Denver with what we could take, or else keep things together by taking some of the wagons on for a few miles and then go back for the rest. The conclusion was to leave four loads of heavy machinery on the plains and go on with the other wagons as fast as possible. I asked the drivers if any of them would stay and guard those to be left. Tobias and the German volunteered to stay.

A GOLD HUNTER'S EXPERIENCE

We selected a camping spot a mile away from the usually traveled road so as to avoid the scrutiny of other pilgrims and look like a small party camping to rest. Then we left them provisions for two or three weeks and went ahead. We guessed that we were then about 150 miles from Denver. The two left behind had no mishaps, but found their stay there all alone for two weeks very dreary and lonesome.

Tobias was for over a year one of my most valuable and agreeable assistants. The German, when in the mountains a short time, lost his eyes by a premature blast of powder in a mining shaft. I helped provide funds to send him East to his friends.

A few days before this misfortune of the death of our oxen and when the drivers were in their most discontented mood, Sollitt, ever suspicious, came to me quite agitated with a tale of gloomy forebodings. He said he had overheard fragments of a talk between the Missourians and some others who were quite friendly with them, which convinced him that a conspiracy was hatching to terminate the tiresome trip, by their deserting us in a body, injuring or driving off the oxen, or committing some more tragic act. He thereupon armed himself heavily with his small weapons, and advised me to do the same.

Instead of following the advice, I became more chatty and friendly with the men and talked of our trials and our better prospects. I discovered in a few a bitter feeling toward Sollitt, occasioned by some rough words or treatment they had received. Sollitt was honest and faithful and in many things very efficient, but was devoid of tact and agreeable ways toward those under his control, especially if he took a dislike to them. One man urged me to assert my reserved authority and take direct charge of the whole business of the train to the exclusion of Sollitt. I had no longings for the disagreeable task of a train master, and simply poured oil on the troubled waters, and went ahead.

When the oxen began to die off, Sollitt told me that he thought one of the Missourians had poisoned them and he disemboweled a number of the dead animals to see if the cause of death could be discovered. He found no signs of poison and nothing that looked suspicious in the stomachs; but he said, the spleens of all of them were in a high state of inflammation. I did not, however, understand that the oxen got their ailment from the Missourians.

One evening we saw the clear cut outline of the Rocky Mountains, including Long's Peak. We differed in opinion, at first, as to whether it was mountain or cloud and could not decide the question till next morning, when, as it was still in view, we knew it was mountain. For several days, though traveling directly toward the mountains, we seemed to get no nearer, which was rather discouraging.

Small flocks of antelope, fleet and graceful, were frequently seen gliding over the plain. They were very shy, and kept several gunshots away. But their curiosity was great, and if a man would lie down on the ground and wave a flag or handkerchief tied to a stick till they noticed it, they would first gaze at it intently and then gradually approach. In this way they were often enticed by hunters to come near enough for a shot.

Forty or fifty miles below Denver we came in view of one picturesque ruin—old Fort St. Vrain—with its high, thick walls of adobe situated on the north side of the Platte. It was built about twenty-five years before, by Ceran St. Vrain, an old trapper and Indian trader.

A GOLD HUNTER'S EXPERIENCE

These adobe walls, standing well preserved in this climate, it seemed to me, would be leveled to the ground by one or two good eastern equinoxial storms.

We reached Denver on the 18th of September about noon, being forty-nine days out from St. Joe. Stubbs met us five or six miles out on the road. This gave him and me a chance, as we walked along, to talk over the condition of things and our plans for the immediate future. He had been in Denver over a week waiting for us and had had no tidings of the train since I wrote him from Fort Kearney. He had considerable liking for display and had evidently told people in Denver that he was waiting for the arrival of a large train of machinery and goods in which he was interested. He thought it would be a scene to be proud of to see fourteen new wagons, heavily loaded and drawn by forty yoke of oxen, come marching into town in one close file.

A GOLD HUNTER'S EXPERIENCE

When he saw only nine wagons straggling along over the space of a mile, covered with dust that had been settling on them for weeks, with oxen lean, footsore, limping and begrimed with sweat and dirt, and teamsters in clothes faded, soiled and ragged, his pride sank to a low level, and he did not want to go into town with the wagons. The train did not tarry, but crossed Cherry Creek—then entirely dry, though often a torrent—drove up the Platte a mile or so and camped for the day on the south or east side of the stream. Stubbs and I spent a couple of hours looking over the town and calling on some acquaintances and then went to the camp.

Denver was at that time a lively place, with a few dozen frame and log buildings, and probably a thousand or more people. Most of them lived and did business in tents and wagons. A Mr. Forrest, whom I had known in

Chicago, was doing a banking business here in a tent. The town seemed to be full of wagons and merchandise, consisting of food, clothing and all kinds of tools and articles used in mining. Many people were preparing to leave for the States, some to spend the winter and to return, others, more discouraged or tired of gold hunting, to stay for good.

When I went to the camp in the afternoon Sollitt and all the drivers wanted to go back to the town to look it over and make a few purchases. I told them I would look after the oxen till evening, when the herders for that night would come and relieve me. The afternoon was clear and warm, though the mountains to the west were carpeted with new-fallen snow. I went out in my shirt sleeves, without a thought of needing a coat. The oxen wandered off quite a distance from camp in search of the best grass, and I leisurely fol-

A GOLD HUNTER'S EXPERIENCE

lowed them. Late in the afternoon, and quite suddenly, the wind sprang up and came directly from the mountains, damp and cold. Soon I was enveloped in a dense fog, and could see but a few yards away. I lost all sense of the direction of the camp or town, and the men at camp did not know where or how to find me. When night came it grew so dark that I could not see my hand a foot from my eyes, and could only keep with the cattle by the noise they made in walking and grazing. Later the fog turned into a cold rain, with considerable wind, and was chilling to the bone, so I was booked for the night in a cold storm without supper or coat. To keep the blood in circulation I would jump and run around in a circle for half an hour at a time. Sometimes I would lean up against one of the quiet old oxen on his leeward side, and thus get some warmth from his body and shelter from the wind. When the oxen had finished grazing and had lain down for the night, I tried to lie down beside one of them to get out of the wind, but the experiment was so novel to the ox that he would get up at once and walk off. During the night the oxen strolled off more than a mile from camp. When morning came I was relieved by the men and was ready for breakfast, and especially for the strong coffee. In times of exposure and extra effort, coffee was the greatest solace we found.

When on a visit to Denver, twenty-three years afterwards, I tried to find out just where I spent that night. An old settler of the place decided with me that it was on the elevated ground now known as Capitol Hill. During the day we crossed the Platte and went forward with the train to the foot of the mountains, and camped some two or three miles south of where Clear creek leaves the foot-hills. Next morning Sollitt took twelve yoke of oxen with two drivers, and started back for the four wagons and two men that had been left behind on the plains. Our teamsters, who had volunteered to drive oxen to the mountains without pay, had now fulfilled their agreement, but most of them were glad to stay with us for awhile at current wages—about a dollar and a half a day. The prospect was not as golden, and the men were not as anxious to get to mining as they had been when a thousand miles further east.

Stubbs had spent a month among the mines and mills, and his observations made him rather blue. The accounts he gave me were most discouraging. He was inclined to think that the best thing for us to do was to go into camp for the winter, look around, watch the developments, and in the spring decide where to locate, if at all, or whether to sell out, give up the enterprise and go home. The proposition was not a bad one, by any means; but I was too full of determination to do *something*, to think of sitting down and quietly waiting six months, after all we had gone through, to get there. I thought we would all be better satisfied if we were to pitch in and make a vigorous effort, even if we failed in the end, rather than to quit at this early stage of the hunt.

The usual route from Denver to the gold fields, was to the north of Clear creek, by Golden City to Blackhawk, and then to Mountain City. Stubbs selected a route further south, because there was a fine camping place, with good grass, about fifteen miles, or half way up to the gold fields, from the foot of the mountains. The roads were quite passable up to this camp, though the hills were steep. With the drivers and oxen that were left after Sollitt started back, the wagons were gradually taken up to this mountain camp, while he was back on the plains and Stubbs and I were looking over the gold region to decide on a final location. The weather was pleasant and rather warm during the day, but frosty at night. We still slept in the open air, and our blankets were often frozen to the ground in the morning.

There was more or less gulch mining and prospecting* going on over a large section of the mountains, but the principal part of the lode mining, and most of the mills that had been located, were confined to a field not over five or six miles in extent, the center of which was Mountain City, now Central City. There were fifty or more mills already up and in running order. They varied

*"Prospecting" included the searching for gold in almost any way that was experimental. Going off into the unexplored mountains to hunt new fields of gold, whether in gulches or lodes was prospecting. Digging a hole down through the dirt and loose stones in the bottom of a gulch to see if gold could be found in the sand was prospecting. Sinking a shaft into the top dirt of a hillside in search of a new lode, or into the lode when discovered to see if gold could be found there was prospecting. And manipulating a specimen of quartz by pulverizing and the use of quicksilver to see if it contained gold was also prospecting.

in capacity from three to twenty stamps. Some were running day and night crushing quartz that was apparently rich in gold; some were running a part of the time, experimenting on a variety of quartz taken out of different lodes and prospect holes, and generally not paying, and some were idle, the owners discouraged, "bust," and trying to sell, or else gone home for the winter to get more money to work with.

The most of these mills were located about Mountain City and Blackhawk and in Nevada and Russell's gulches. The rest of them were scattered in other small gulches or mountain valleys in the vicinity. The richest mines being worked were the Bobtail, Gregory, and others, in Gregory gulch between Mountain City and Blackhawk. The other principal gold diggings were some seventy miles further south, near the present site of Leadville. These I did not then visit. Nearly all of these mills had been brought out and located during the year 1860. Ours was about the last one to arrive that season. It was evident that the business was not generally paying. The reasons given were, that the mills did not save the gold that was in the quartz, and that those at work in the mines were nearly all in the "cap rock" which was supposed to overlie the richer deposits below. The theory was that the deeper they went the richer the quartz. There were just enough rich "pockets" and streaks being discovered and good runs made by the few paying mines and mills to keep everybody hopeful and in expectation that fortune would soon favor them. So they worked away as long as they had anything to eat, or tools and powder to work with.

After looking over the fields a number of days, carrying our blankets and sleeping in empty miners' cabins, Stubbs and I concluded to locate at the head of Leavenworth gulch, which was about a mile and a half southwest of Mountain City, between Nevada and Russell's gulches. The side hills were studded all over with prospect holes and mining shafts. Several lodes, said to be rich in gold, had recently been discovered, and a nice stream of water ran down the gulch. Only three mills were in operation there, and a number of miners who were developing their own claims strongly encouraged us to come, promising us plenty of quartz to crush. Several parties were gulch mining there with apparent success, and during the short time that I watched one man washing out the dirt and gravel from the bottom of the gulch he picked up several nice nuggets of shining gold, which was quite stimulating to one's hopes. I afterwards learned that these same nuggets had been washed out several times before, whenever a "tenderfoot" would come along, who it was thought might want to buy a rich claim.

As soon as we located and selected a mill site, we went vigorously to work, and all was preparation, bustle and activity. Stubbs was a good mechanic and took charge of the construction. Others were cutting down trees, hauling and squaring logs, and framing and placing timbers to support the heavy mill machinery. As soon as Sollitt returned from the plains, he, with a few of the drivers, went to work to get the wagons, machinery and provisions from the mountain camp up to our location. In many places, at first glance, the roads looked impassable. They went up hills and rocky ledges so steep that six yoke of oxen could pull only a part of a load; then down a mountain side so precipitous that the four wheels of each wagon would have to be dead-locked with chains to keep them from overrunning the oxen; then they would go along mountain streams full of rocks and bowlders, and upsetting a wagon was quite a common occurrence. I saw one of our provision wagons turn over into a running stream, and, among other things, a barrel of sugar start rolling down with the current.

As soon as everything was brought up to our final location, I sold some of the wagons, some oxen and the pony, thus securing cash to pay help and other expenses. I traded others off for sawed lumber, shingles, etc., for use in building the mill-house and a cabin. Grass was very scarce in the mining regions. One of the faithful, well-whipped oxen was killed for beef (a little like eating one of the family). In this dry, pure air the meat kept in perfect condition for many weeks till all eaten up, and it was an agreeable change in our diet.

When we had finished the hauling of timber and other things, we sent the oxen, still on hand, down to the foot of the mountains where there was grass during the winter; for cattle would pick up a living among the foot-hills, and come out in good condition in the spring. The distance was some twenty-five or thirty miles. Early one bright November morning I started down there on foot to make arrangements with a ranchman to look after them. The air was so bracing and stimulating to the energies that I felt as if a fifty-mile walk would be mere recreation. Being mostly down hill, I arrived at the ranch before noon, did my business, got a dinner of beef, bread and coffee, and felt so fine that soon after two o'clock I concluded to start for home, thinking that in any event I would reach one of the two or three cabins that would be found on the latter part of the road. Walking up the mountains was slower business than going down, and long before I reached the expected cabins it became dark and I was completely tired out. I found a small pile of dried grass by the roadside which had been collected by some teamster for his horses. I covered myself up with this as well as I could, and being very tired, was soon asleep, without supper or blanket. On awakening in the morning, I found myself covered with several inches of snow, and felt tired, hungry and depressed. I plodded along toward home for a few hours, and came to a cabin occupied by a lone prospector, who got me up a meal of coffee, tough beef and wheat flour bread, baked in a frying pan with a tin cover over it. Soon after finishing the meal I felt sick and very weak, and was unable to proceed on my journey till late in the afternoon, when I went ahead and reached home long after dark.

Leavenworth gulch was crossed by dozens of lodes of gold-bearing quartz, generally running in a north-easterly and south-westerly direction. In this district the discoverer of a lode was entitled to claim and stake off 200 feet in length, then others could in succession take 100 feet each, in either direction from the discovery hole, and these claims, in order to be valid, were all recorded in the record office of the district. Owners of these various claims, to prospect and develop them, had dug the side hills of the gulch all over with hundreds of holes from ten to thirty feet deep, partly through top dirt and partly through rock. A few would find ore rich enough to excite and encourage all the rest. More would find rich indications that would stimulate them to work on as long as they had provisions or credit to enable them to go ahead, hoping each day for the golden "strike." A large majority of these prospect holes came to nothing. Many of the miners had claims on several different lodes, and although they might have faith in

A GOLD HUNTER'S EXPERIENCE

their richness, they wanted to sell part of them to get means to work the rest. We had plenty of chances to buy for a few hundred dollars in money or trade mines partly opened, showing narrow streaks of good ore, which, according to the prevailing belief, would widen out and pay richly as soon as they were down through the "cap rock."

While work was progressing on the mill I spent considerable time in looking over these mines, and I went down numerous shafts by means of a rope and windlass, turned by a lone stranger, who I sometimes feared might let me drop. I listened to glowing descriptions by the owners, examined the crevises and pay streaks, and took specimens home to prospect. This was done by pounding a piece of ore to powder in a little hand mortar, then putting in a drop of quicksilver to pick up the gold, and then evaporating that fluid by holding it in an iron ladle over a fire. The

A GOLD HUNTER'S EXPERIENCE

richness of the color left in the cup would indicate the amount of gold in the quartz.* I could soon talk glibly of "blossom rock," "pay streaks," "cap rock," "wall rock," "rich color," and use the common terms of miners. I bought two or three mines, traded oxen and wagons for two or three more, and furnished "grub stakes" to one or two miners—that is, gave them provisions to live on while they worked their claims on terms of sharing the results.

Quartz mills were nearly all run by steam and the fuel was pine wood cut from the mountain sides, every one taking from these public domains whatever he wanted. The principal features of our mill were twelve large pestles or stamps, weighing 500 pounds each, which were raised up about eighteen

---

* In testing quartz by specimens, "greenhorns" were sometimes deceived by "loaded" quicksilver, that is by that which had some gold in it and would leave a "color" whenever evaporated. I knew one miner who worked away in his mine, taking out quartz all winter, and was in good spirits as he tested a specimen of his ore every day or two and always found a rich color. When crushed in the spring his quartz did not "pay." The bottle of quicksilver he had used all winter was found to be "loaded."

inches by machinery and dropped into huge iron mortars onto the small pieces of rock which were constantly fed into them by a man with a shovel. A small stream of water was let into the mortars, and as the rock was crushed into fine sand and powder it went out with the water, through fine screens in front, and passed over long tables, a little inclined, and then over woolen blankets. The tables were covered with large sheets of brightly polished copper. On these polished plates, quicksilver was sprinkled and it was held to the copper by the affinity of the two metals for each other. As the water and powdered rock passed over the tables, the quicksilver, by reason of its chemical attraction for gold, would gather up the fine particles of that metal and, as the two combined, would gradually harden and form an amalgam, somewhat resembling lead. Coarser grains of gold would lodge in the blankets, owing to their weight,

A GOLD HUNTER'S EXPERIENCE

while the small particles of rock would pass over with the water. The amalgam was put into a retort and heated over a fire, when the quicksilver would pass off in vapor through a tube into a vessel of water, and then condense, to be again used, while the gold would be left in the retort, to be broken up into small pieces and used as current money. In order to save as much of the gold as possible, these copper plates required close watching, constant care and much rubbing to remove the verdigris that would form.

About the first of November our mill was completed, and we expected to operate it a good part of the winter with the quartz of other miners, together with that which we would take out ourselves from our own mines. A large well, or underground cistern, was dug under the mill house, which was fed by copious springs, and promised to furnish an abundant supply of water. To furnish water for the numerous mills about Mountain City and in Nevada gulch a large ditch had been dug, which started up in the mountains near the Snowy range, and wound like a huge serpent around promontories and the sides and heads of numerous gulches, with a slight incline, for some fifteen miles. It passed around the hills which bordered Leavenworth gulch, a few hundred yards above our mill site. About the time the mill was completed the water was turned off from this ditch on account of freezing weather and the near approach of winter. Very soon after, the beautiful springs which supplied our tank and the gulch with water, all dried up. They had been fed by seepage from the big ditch. With the disappearance of the water vanished all prospect of running the mill before spring, when the melting snow would furnish a supply. It seemed like a bad case of "hope deferred." But the bracing air and climate, outdoor life, constant exercise, coarse food and pure water were too invigorating and stimulating to the feelings and hopes to allow one to feel much depressed or discouraged. We looked forward to the next summer for the golden harvest.

Stubbs built us a one-and-a-half-story-cottage out of sawed lumber, boards and shingles, with one room below for living, eating, cooking and storing provisions in, and one above for a dormitory. A corner of the latter was partitioned off into a small room for him and me, with a bunk for each, under which we stored our twelve kegs of powder, as being the safest place we had for it. We slept on beds of hay with our blankets over us, and in very cold weather piled on our entire stock of coats and some empty provision sacks. In the room below was a good cook stove, and there was wood in abundance, so we kept comfortable, though the house was neither plastered nor sheeted, and considerable daylight came in through cracks in the siding. We had a table and benches made of boards, and Stubbs made me an armchair and a desk for my account books, papers and stationery. What a luxury, after four months camping out, to be able to sit down in a chair, eat from a table, sleep on a bed, write at a desk, read by a candle at night and have regular, well-cooked meals.

To a lover of the picturesque in scenery our location was ideal. Immediately around us was a semicircle of high, steep, pine-covered hills spotted with prospect holes. To the east, through an opening in the intervening mountain ranges, the plains were in full view over a hundred miles away. Sometimes for days, they were covered with shifting clouds which seemed far below us. Then an east wind would drive the clouds and mist slowly up into the mountains, swallowing up first one range and then another, till only a few peaks would stand out, above an ocean of fog, and finally we would be enveloped ourselves. Ascending a hill a few hundred yards above our house and looking westward over a great depression or mountain valley, one had in full view the Snowy range over twenty miles away, with its crests and peaks covered with perpetual snow, and Mount Gray still further in the distance. In the fall and winter almost every day local snowstorms and blizzards were seen playing over this great basin and on the sides of the distant range. Our location was some nine or ten thousand feet above the sea. The lightness of the air gave some inconvenience and many surprises to new comers. They would get out of breath in a few minutes in walking up a hill. I would wake up several times in a night with a feeling of suffocation, draw deep breaths for a few minutes and thus get relief before going to sleep again. It took ten minutes to boil eggs, two to three hours for potatoes, and beans for dinner were usually put on the fire at supper time the day before.

Coin and bank bills were seldom seen. The universal currency was retorted gold, broken up into small pieces, which went at $16 an ounce. Every man had his buckskin purse tied with a string, to carry his "dust" in, and every store and house had its small scales, with weights from a few grains to an ounce, to weigh out the price when any article from a newspaper to a wagon was purchased. No laws were in force or observed except miners' laws made by the people of the different districts. When a few dozen miners, more or less, settled or went to work in a new place they soon organized, adopted a set of laws and elected officers, usually a president, secretary, recorder of claims, justice of the peace and a sheriff or constable. Appeals from the justice, disputes of importance over mining claims, and criminal cases were tried at a meeting of the miners of the district. We were in the district of Russell's gulch. Sometimes we had a meeting of the residents of our own gulch. One chap there stole a suit of clothes. The residents were notified to meet at once, and the same day the culprit was tried and found guilty, and a committee, of which I was one, was appointed to notify him to leave our locality within two hours and not to return, on penalty of death. He went on time. Had he been stubborn and refused to go, I don't know what course the committee would have taken. This member of it would have been embarrassed. An adjoining district was made up mostly of Georgians. They had their own tastes and preju-

A GOLD HUNTER'S EXPERIENCE

dices. Soon after we came to the mountains, at their miners' meeting a man was convicted for some offence and sentenced to receive thirty lashes from a heavy horsewhip. The day for the execution of the sentence was regarded as a kind of holiday and the miners collected from all the country around. All our men, including Sollitt, went to the whipping. Stubbs and I stayed at home. We had no relish for that sort of amusement. A thief was more sure of punishment than a murderer. There was so much property lying around in cabins unguarded, while the owners were off mining or prospecting, that stealing could not be tolerated, while the loss of a man now and then by killing or otherwise did not count for much.

When it was found that the mill could not be run during the winter, we discharged all the men except the cook, and two others, who were kept to help

A GOLD HUNTER'S EXPERIENCE

do a little mining on two of the claims that we had secured by trade and purchase. A shaft about three feet by six was sunk in each, which followed the vein of mineral quartz down to a depth of thirty to fifty feet. In one, the vein was quite rich in places, but only two or three inches wide, and it would not pay to work it; but the hope that kept us, like hundreds of others at work, was, that the vein would widen out when we got a little deeper and grow richer as it went down. This hope was never realized. The other shaft was on a lode called the Keystone, and developed a wide vein of black pyrites of iron that much resembled that which was being taken out of the best paying mines, and most of the miners that examined it declared that we had a bonanza. Of course we were in good spirits, but we did not care to run in debt in order to take out more mineral than we got in sinking the shaft, of which there were several cords. I worked a part of each day in the shafts, with the others, to learn the details, drilling, blasting and picking out the "pay streak." Then I spent a good deal of time looking around among other mines, and the mills that were at work, to learn what I could. Quite a number of other miners were at work in the gulch sinking shafts on their best claims and taking out ore to be crushed in the spring. To some of these we furnished provisions to enable them to keep at work. Most of the roving, restless, fickle people had gone home in the fall and those who stayed were men of grit and determination. Some of them were well educated and intelligent. Every little while somebody would strike a small pocket, or a streak of very rich ore, which would help to make everybody else feel hopeful. And so the winter wore away.

There were four families in the gulch this winter, including that number of women, several children and three young ladies. The young men buzzed around the homes of the latter like bees about a honey dish. These families united and had a party on Christmas Eve. Three cottages were used for the occasion, one to receive the guests in, ours for the supper room, and another with a floor for dancing. We regarded this as the "coming out" of the youngest of the young ladies. Several ladies from Russell's and other gulches came to the party. Among those living here were quite a number who brought a few books with them. No one person had many, but all together they made quite a library and were freely lent. I remember borrowing and reading by the light of a candle, in these long winter evenings, some works on mines, Carlyle's works, a few histories and several novels. The almost universal amusement with the miners and others was card playing, confined to euchre and poker. Every miner had a pack of cards in his cabin if not in his pocket, and generally so soiled and greasy that one could not tell the jack from the king. Gambling was common and open in Denver and Mountain City, and not unusual elsewhere. Playing for gain was never practiced in our cottage. When poker was played, beans were put in the jackpot instead of money.

Near the junction of Russell's and Leavenworth gulches, and about a third of a mile from our location, was a mill owned and run by George M. Pullman, then a comparatively obscure man, but later known to the world as the great sleeping car magnate. He also had an interest in a general supply store near Mountain City. He lived much of this winter in a cabin near the mill, and rode back and forth to town almost daily on an old mule. He wore common clothes like the rest of us, and the only sign of greater importance that he exhibited was, that while I walked to town, he rode the mule. He left the mountains the next summer for Chicago, and entered upon his sleeping-car enterprise, which led to fame and fortune.

Another young miner that was much in evidence about Mountain City this winter was Jerome B. Chaffee, who afterwards made a fortune in mines, took an active interest in local politics and became a United States Senator.

In Mountain City there was an enterprising chap who started a pie bakery and did an extensive business. Miners from all the country around, when they came to town, crowded his shop for a delightful change from the usual cabin fare. I went to town every few days for letters and papers, or to visit the mills, and always indulged in this one dissipation. I went to his bakery and feasted on pie. He had peach, apple, mince, berry, pumpkin and custard pie, and never since I was a boy in the land of pie did the article taste so good.

Within a hundred yards of our mill lived and worked the gulch blacksmith, named Switzer. He sharpened our drills and did our smith work generally. He had a bitter feud with a gambler in Mountain City, which resulted in each vowing to shoot the other on sight. They carried loaded revolvers for the occasion for nearly a month, and then happened to meet in broad daylight in the principal street of the town. The other fellow was the quicker—Switzer fell dead and we had to find another blacksmith. No notice was taken of the affair by the authorities.

Sollitt became ill with what the doctors pronounced scurvy, and went East before April. Stubbs and he disliked each other from the first, and whatever one suggested the other opposed. This made it easier for me to decide some questions, as I never had both of them against me. The people here were generally very healthy. I increased much in strength and vigor, and weighed 175 pounds for the first and only time in my life. November was windy, stormy and cold, but in December the weather was settled and pleasant. During the winter the mercury a few times went below zero; otherwise the climate was delightful. The warm sunshine of the last half of April melted the snow, thawed the ground and brought a supply of water for the mill, even before the big ditch began to run. We soon began crushing the piles of quartz that had been taken out during the winter by various miners, and tried our own rich-looking black stuff from the Keystone. The mill was run day and night. I took charge from midnight till noon and Stubbs from noon till midnight. None of the rock was found rich enough to pay for mining and milling. That tried in one or two other mills was no better. General discouragement followed, and everybody stopped mining in our gulch. Some went to work for wages in other mines, to get a fresh supply of provisions, etc. Some went off prospecting and gulch mining in the newer gold regions. Our neighbor, Farren, moved his mill seventy miles away, to California gulch, near where Leadville now is. A mill partly erected near our mill site, and owned by a Mr. Bradley and a Mr. H. H. Honore, the father of Mrs. Potter Palmer, was moved away to other parts, and our mill was left alone. The gulch was soon almost deserted. Mines and mills seemed to be of no use or value. Our whole enterprise had apparently collapsed, and the golden halo, that for ten months had surrounded it, had vanished. Hope departed, and for a few days was replaced by feelings of disappointment and depression of spirits

A GOLD HUNTER'S EXPERIENCE

not often experienced by me. Stubbs abandoned the business and decided to go home and leave me to hold the fort and look after the wreck, as he called it, to see what could be saved.

He built a boat, had it hauled down to the Platte at Denver, piled in his provisions and effects, launched it in the river and started down stream, hoping to reach Omaha in that way. All went well for about a hundred miles, when the water grew so shallow that he was stranded amid the small islands and shifting sands. He got ashore, abandoned his boat and took passage in an eastward-bound mule wagon. He and the principal, Mr. Sollitt, afterwards sold out their interest in the enterprise to Mr. Ayres for a small consideration.

In a few days I got over the "dumps," and spent a week or two visiting the newer gold fields up the south branch of Clear creek, about

Idaho, Georgetown, Empire and Fall river, where new lodes were being discovered almost daily. Not much gold was being taken out, but everybody was full of hope and expectation and busy prospecting and staking off claims on newly discovered lodes. I had some staked off for myself by some men who had worked for us.

Geo. M. Pullman wanted to experiment on a load of the ore from our noted Keystone lode, as it looked so rich. When it was going through the mill, the amalgam piled up so fast on the copper plates and appeared so rich that he at once came up to see me and proposed that we buy, on joint account, the adjoining claim on the same lode, as I knew the owner and had formerly had an option on its purchase. A few hours later, when they had cleaned up and retorted the amalgam he came galloping up again on the old mule to stop proceedings, as they got very

A GOLD HUNTER'S EXPERIENCE

little of value from the amalgam, and that mostly silver. Thus that gleam of hope quickly vanished also.

Late in June, with Tobias as a companion, I took a trip of observation over the range into the wild regions of Middle park. We carried our blankets, flour, bacon, coffee and sugar to last a week, also tin cups, plates and spoons, a frying pan, gun, pistol, hatchet and belt knives. Walking the first day slowly up the slopes through the pine forests, around the head of Nevada gulch, and along the high ridge south of Boulder valley, we camped for the night just below the timber line so as to have fuel for a fire. A few tracks of Mountain lion were seen in the afternoon. The trees grew smaller and smaller till the last seen were old ones covered with moss and only a few feet high. After leaving the line of timber growth, the ground for some miles was thickly carpeted with mountain moss, then in full bloom in rich colors of red, white, blue and yellow. In the afternoon we reached the top of a high peak on the crest of the range where all was desolation, and nothing grew. The peak was a vast pile of broken rocks and stones partly covered with snow. To the North Long's Peak stood out above everything else. To the East one had a grand view over a wilderness of mountain ranges and peaks to the great plains in the dim distance. To the South, beyond a range of other snow-capped peaks, towered Mount Gray. Within a mile of us in full view, were seven mountain lakes from ten to a hundred acres in size, and one of them, which was screened from the sun's rays by a steep rocky ledge, was still solid ice from the freeze of the last winter. To the west was visible a circle of mountain tops, thirty or forty miles away, and surrounding the great basin, a mile below us in elevation, which constituted Middle park. The afternoon was bright and pleasant, and we decided to spend the night on the peak, to see the sunrise and enjoy the view in the clear morning air. We made a bed with flat stones and rolled up in our blankets for sleep. Then the wind blew over us and up through the crevices in the rocks under us and soon our teeth were chattering and we were chilled through and through. To keep from freezing we climbed in the darkness, over the rocks and down the mountain side to a sheltered nook, then rolled up and went to sleep. During the night I was awakened by some animal sniffing about my head and pulling at my blanket. A yell, a start and two or three stones thrown after him, sent him off among the rocks, and I never knew what it was. At daylight we again climbed up the peak, saw the sun rise, made a breakfast of bread and sugar as we had no fuel to make a fire, and then started down the mountain. The little streams and pools coming from the melting snows the day before were now all frozen up.

By ten o'clock we were down where the vegetation was luxuriant, the flowers in bloom and the butterflies flitting about them. Along the stream that we descended to the westward, was a series of beaver dams continuing for several miles, covering two or three acres each, with breasts four or five feet high formed of logs and brush. Out in the middle of the dams were the beavers' houses, partly under water and rising a few feet above. Many of the logs, cut off by the beavers to form the dams, and the stumps on the shore where they had gnawed down the trees, were twelve to fifteen inches through. Further on we saw bear tracks in the mud along the stream. When we camped at night we made a bed of pine boughs, and over it a small shelter with branches of trees cut with the hatchet. We built a fire on the side hill above our sleeping place beside a fallen tree. In the night it burned through and a log rolled down the hill over us, and we awoke with a sudden start. I thought of bears and instantly seized my hatchet and knife for defense, before realizing the true situation. Old skulls and bones of buffalo were plentiful, showing that the animals had once occupied these fertile valleys. On starting back we followed an old animal trail, the general course of which was headed toward the range, though it wound around the mountain sides and gulches in all directions. We felt sure it would lead over the Snowy range at the easiest passage. After following it two days, often climbing over and creeping under fallen trees, it brought us through a low pass to the head waters of South Clear creek, whence we had an easy trail down hill most of the way home.

Though far away from the seat of the civil war we did not escape its excitements. The Southerners were numerous in the mountains, and of course all sided with the South. They and the Northerners were very suspicious of each other, and each party bought up all the guns they could get in the mountains. During the summer of 1861 much fear was felt that a rebel force might march up the Arkansas and, with the help of their friends here, capture the whole settlement. But when the Southern troops were defeated and driven out of New Mexico by the Union forces in the following spring, all danger was over and "Pike's Peak" was loyal. The Southerners gradually left to join the rebel army. We got news from the East in six days, by telegraph to Omaha, the overland mail coach to Julesburg, near the forks of

A GOLD HUNTER'S EXPERIENCE

the Platte, and by pony express from there to Denver. St. Louis papers were eight days old and Chicago papers ten days old when received.

One of the best known miners in our region was Joe Watson, who came from near Philadelphia, in 1859, and he came to stay. Though quiet and unassuming he was nervy, determined, persevering and persistent. He discovered, staked off, owned and worked many claims in Leavenworth and other gulches. Sometimes he had streaks of luck and often the reverse. When lucky he would hire men to help him, when "broke" he would put more patches on his clothes, sharpen his own tools, borrow a sack of flour and work away. Some years later he discovered a really rich gold mine, then worked a silver mine in Utah and became a millionaire. During the spring of 1861 and the winter previous, he prospected in several of his claims, but fortune was

A GOLD HUNTER'S EXPERIENCE

against him. In July, when most of the other miners had left our gulch, he came back and quietly went to work in a claim that he owned on the hillside a few hundred feet above our cottage. In two or three weeks he took out from a narrow crevice two cart loads of top quartz which looked like rusty iron (not having got down to the pyrites), and he persuaded me to start up the mill and crush it. Very soon the amalgam began to pile up on the copper plates as I had never before seen it. The result of the "clean up" and retorting was $1,000 worth of shining gold. The next run, out of the same mine, produced but little gold, a good example of how that metal was found in streaks and pockets. Watson paid his debts, got a new suit of clothes, laid in a stock of provisions, and went to work again developing his mines. It was related of him that he went to Philadelphia one winter to try and sell shares in his mines, and that he wore a suit of Quaker clothes, used the plain language, attended Friends' meetings, and had good success in selling shares. Of these early workers I might name a few more who attained wealth or prominence; but the great majority — those who hoped and struggled and toiled without success, are forgotten.

The rich strike in Joe's mine made quite an excitement. Some others were inspired with renewed hopes and many visited the gulch to see the rich mine they had heard of. There was a small army of miners marching through the mountains constantly, going in all directions, leaving one place for some other where rich strikes were reported.

I concluded to make one more trial in the Keystone, dig a little deeper and see if the ore was any richer there. The result was a pleasant surprise, and gold enough to more than pay expenses. I hired a gang of men to work the mine night and day, and thus kept the mill going till the water gave out in the fall. As I had no skilled assistant I had to work at least sixteen hours a day in running the mill, procurring supplies and superintending everything. Some runs proved the quartz to be quite rich, though it varied greatly. We still believed in the theory that it would grow richer as we went deeper. I arranged to mine all winter and pile up the quartz for spring crushing.

In April, 1862, when provisions were nearly used up in the mountains and the early spring supply trains from the East were about due, there came an unusual fall of snow, eighteen inches deep, extending far eastward over the plains, completely blockading teams and transportation. A famine was threatened and people became panic-stricken. Flour rose as high as $50 a sack, and one day a small quantity sold for eighty cents a pound. Coffee and

A GOLD HUNTER'S EXPERIENCE

other things also advanced in price. We were on our last sack of flour, and I decided that when that was gone the men must all quit work and start eastward to meet the supplies on the plains. But the incoming trains soon began to arrive in Denver, and provisions were plentiful at usual prices.

When the mill was started up in the spring our hopes were dashed by finding that the quartz taken out during the winter did not pay as well as that of the previous season. The mine was down about a hundred feet, and the last taken out did not pay expenses, so I discharged the miners again. I was getting tired and disgusted with the whole business, and realized that it was about time to return East if I were going back there to settle down.

About the first of June, Mr. Ayres came out to spend the summer. He was so delighted with the beauty of the scenery and novelty of the business that he talked of sending for his family. The mountain sides were gay with wild flowers in full bloom in gorgeous colors. The shining gold that he could see taken out by several successful plants, delighted his eyes and stimulated his imagination nearly up to the point of genuine gold fever. His coming was of course a great relief to me by dividing the responsibility and work about the mill. We ran the mill night and day, crushed all the quartz that could be got and worked over a large pile of tailings that had accumulated below the mill, which paid a small profit. The summer's success was very moderate. About midsummer Mr. Ayres bought out my interest in the enterprise, with the understanding that I would remain till fall and assist him. He wanted to give the business a further trial. I determined to return to Chicago and try to take advantage of the tide of prosperity then beginning to rise in the East.

Mr. Ayres remained till late in the fall, then went to Chicago for the winter and returned to the mountains early in the spring of 1863, to give the business a further trial. But he did not do much mining or milling. During that spring and the following summer a fever of speculation prevailed all over the East, brought about by the war and the deluge of greenbacks. It extended to mining stocks, and especially to gold mines, as gold was then selling at a high premium — one hundred dollars in gold bringing $260 in legal tender currency. Mr. Ayres offered his plant for sale, went to New York in the summer and disposed of it in Wall street for $30,000. The mill was never afterwards run and I believe, none of the mines ever worked. Twenty years later I visited Leavenworth gulch. The mill and all the houses and cabins of my former days there had disappeared, and most of the old prospect holes and mining shafts had caved in. One familiar sight, however, remained. A load or so of black, rich looking ore was lying upon the ground unused and uncared for at the shaft of the Keystone.

On the 22nd of October, 1862, I left the mountains and gave up the mining business for ever. The next day at Denver I took passage for Omaha, in a two-horse covered wagon, with a man and his wife who were returning to their home in Baraboo, Wis., after spending two years in the gold fields with only moderate success. Another man also took passage making a party of four. Leaving the wagon to the man and his wife, my fellow passenger and I slept on the ground in our blankets, except occasionally, when near some ranch or settlement, we could enjoy the luxury of a haystack. When two or three days out of Denver we had a "cold snap" which froze the vegetables in the wagon and made sleeping out very uncomfortable. The woman did the cooking and the men collected the fuel. The other two men had guns and supplied us with small game. We saw a few dozen buffalo, but they were too far off to shoot. One day the two men went off on an all-day hunt among the distant hills, the arrangement being to meet us in camp at evening. I drove the team, and in the afternoon we came in sight of a camp of Indians with their lodges set up near our trail. The only thing to do was to drive boldly ahead. The woman sat on a seat well back in the wagon, and I sat forward with my feet out on a front step. I hung up a blanket close behind me across the wagon, so that the Indians could not see how many persons were in it. As we approached the camp about a dozen of them came out on the trail in front of us, motioning to me to stop and calling out, "Swap, swap, swap," meaning for us to stop and trade with them, but intending doubtless to find out how many were in the wagon, and rob us if they dared. Suddenly, when within a few yards of them, I whipped the horses with all my might, and drove furiously past and away from the camp. When our party met at night, all agreed that the day's experience savored too much of danger to allow the hunters to go out of sight of the wagon again.

We passed two or three camps of Sioux Indians along the Platte, but they gave us no trouble. When driving through the trees and bushes in a lonely spot about a day's journey below Fort Kearney, we suddenly met a band of mounted Pawnee warriors, who stopped us and in broken English asked where we were going, where we came from, if we saw any Sioux Indians, how big the

A GOLD HUNTER'S EXPERIENCE

bands were, if they had many ponies and how many days' journey they were away. We answered their inquiries, and they told us to go ahead. They rode westward, doubtless to make a raid on their enemies, the Sioux.

The weather was now getting cold; we approached the settlements and enjoyed the haystacks. One night, while camping near an Indian settlement on the Platte, I crawled well into the middle of a small rick of hay. The Indians were tramping around it and over it and howling and yelling all night, but I kept my berth till morning. We reached Omaha in twenty days from Denver. There I said good-by to my traveling companions and took stage for Iowa City, whence I could go by rail to Chicago. The stage trip was two days and nights of continuous travel, except short stops to change horses and get something to eat. We were packed three on a seat, with no chance to stretch out our limbs, and no opportunity for sleep, except such as could be obtained sitting upright and jolting over the rough roads.

After an absence of about two and a third years, I reached Chicago in the middle of November, 1862, a wiser if not a richer man.

After selling out my interest in the joint enterprise, I still had left some fifty claims on various lodes in the newer gold fields of the Clear creek region. Some I had pre-empted, and some I had bought in job lots from miners who were "broke" or were about to leave the mountains. Some had prospect holes dug in them and some were entirely undeveloped. They may have been worthless, and they may have contained untold millions. But I had given up the mining business. Some time after returning to Chicago I was making a real estate trade, and we were a little slow in adjusting the difference in values and closing the deal, and finally as "boot" to make things even I threw in these fifty gold mines. Perhaps this was a mistake and a squandering of wealth and opportunities. Had I only kept them, and gotten up some artistic deeds of conveyance, in gilded letters, what magnificent wedding presents they would have made. And the supply would have been as exhaustless as that of Queen Victoria's India shawls. In the long list of high-sounding, useless presents, the present of a gold mine would have led all the rest.

In summing up the losses and gains of the expedition, I have to charge on one side two years and four months of time devoted to hard work, with many privations, and about $500 in cash which I was out of pocket. On the other side, I had built up a fine constitution, increased in strength and endurance, gained valuable business

A GOLD HUNTER'S EXPERIENCE

experience, learned in a measure to persevere under difficulties, and to bear with patience and fortitude the backsets, reverses and disappointments that so often beset us, and, finally, had learned enough not to be taken in by the schemers who are constantly enticing eastern people to invest in gold and silver mines. Did the enterprise pay?

INDEX

A

Adobe 27, 59, 60
Alkali plains 54
Antelope 59
Arkansas 33, 102
Ayres, Enos 6, 9, 95,
  107, 108, 109

B

Bank in a tent 62
Baraboo, WI 110
Bears 101
Beaver dams 100
Big Blue River 20
Blackhawk mine 66, 68
Blacksmith 92
Bobtail mine 68
Boulder Creek, CO 6
Boulder Valley 97
Bradley, Mr. 94
Buffalo 22, 23, 30-32,
  33, 34, 36, 40ff, 40-41, 50,
  111;
  skeletons 39, 101
Buffalo skins 47

"Buffalo chips" 42

C

Cactus 51
California 7, 28
  mail to 27
California Gulch 94
"Cap rock" 69, 76, 77
Capitol Hill 64
Carlyle 89
Cattle 16, 34
Cattle thieves 35
Chaffee, Jerome B. 91
Cherry Creek 61
Chicago 5, 6, 8,
  9, 10, 14, 34, 62, 91, 103,
  108, 109, 113, 114
Cholera 6
Christmas Eve 89
Civil War 102
Clear Creek . . 64, 66, 95, 114
Cook 14, 34, 86
Copper plates . . . 78, 79, 104
Cottage 81
Cottonwood 29
Custer, General 40ff

A GOLD HUNTER'S EXPERIENCE

D

Denver, CO 6, 29, 55, 56, 59, 60, 61, 64, 66, 95, 103, 107, 110, 111, 113
Dentist 10
Depression 5
Dogs 47

E

Empire 96

F

Fall River 96
Farren, Mr. 94
Forrest, Mr. 61
Fort Harker 40ff
Fort Hays 40ff
Fort Kearney 22, 27, 29, 30, 32, 41, 60, 112
Fort Leavenworth 20
Fort St. Vrain 59
"Friends" [Society of] 105
Frogs 24

G

Gambling 90
Georgetown 96
Georgian prospectors .. 5, 85
Germans 10, 55, 56

Gold

Gold 3, 5, 44, 67ff, 68-69, 70, 74, 76, 77, 78, 79, 84, 96, 103, 104, 108, 109, 116
Gold fields 66, 95
Gold mines 114-115
Golden City 66
Graves 54
Gregory mine 68

H

Hiawatha, KS 17
Honore, Mr. H.H. 94
Horsewhipping, public ... 86

I

Idaho 96
Illinois 10
Indians 41, 45, 111
  dressing skins 49; lodges 47
Inman, Henry 40ff
Iowa City 113

J

Julesburg 102

K

Kansas 8, 17
Kansas Pacific R.R. 40ff

A GOLD HUNTER'S EXPERIENCE

Keystone lode 87, 93, 96, 105, 110

L

Lariat, rawhide 11
Leadville 68, 94
Leavenworth Gulch .. 70, 74, 80, 90, 103, 109
Library, lending 89
Little Blue River .. 20, 22, 24
Locomotive boiler 48
Long's Peak 58
Lumber 72

M

Mail coach 102
Mail, U.S. 27
Marysville 20, 21
Mexican saddle 11
Middle park 97, 99
Mines 6, 67, 69
Missouri 10, 28, 57, 58
Missouri River 8
Moccasins 49
Mortar, hand 76
Mortar, iron 78
Mountain City ... 66, 67, 68, 80, 90, 92
Mountain City Mine 68
Mountain lion 99
Mount Gray 83, 98

Murder 86

N

New Mexico 102
North Long's Peak 98

O

Old Santa Fe Trail 40ff
Omaha 95, 102, 113
Oregon 28
Oregon Trail 20
Owls 52
Oxen .... 9, 11, 12, 18, 24, 37, 42, 43, 52-53, 62, 63, 65, 71, 72, 73; death of 13, 56, 58

P

Palmer, Mrs. Potter 94
Panic of 1857 5
Pawnee Indians 26, 112
Philadelphia 103, 104
Pike's Peak 5, 11, 28, 43, 102
Pike's Peak Expedition 3
Pike's Peak gold fields 39
"Pike's Peak or Bust" 45
Platte River 25, 27, 29, 33, 42, 61, 64, 95, 112, 113
Platte, Forks of 103

119

Platte Valley 40
Pony 17, 22, 36, 46, 47, 50, 72
Pony Express 103
Post office 27
Prairie dogs 51
Presbyterian 10
Prospecting 67ff
Pullman, George M. ... 90, 96

Q

Quaker clothes 105
Quartz 68, 69, 70, 74, 77ff, 79, 93, 106, 108
Quartz mills 7, 77
Queen Victoria's India Shawls 115
Quicksilver 7, 76, 78, 79

R

Rail travel 113
Rattlesnake 35, 52
Rocky Mountains 58
Russell's Gulch 68, 70, 89, 90

S

Sagebrush 26
St. Joe, MO 8, 10, 14, 41, 60

St. Joe River 27
St. Louis 40ff, 103
St. Vrain, Ceran 59
Salt Lake 28
Scurvy 92
Sheridan, General 40ff
Silver 97, 103
Sioux Indians ... 46, 112-113
Snakes 24
Snow 74
Snowy Range 80, 101
Sollitt, John 6, 7, 8, 9, 12, 13, 21, 22, 23, 26, 31, 35, 56, 57-58, 62, 64, 66, 71, 86, 92, 95
South Cedar Creek 101
Stagecoach 8, 40ff, 113
Stampede 38
Stamp mill 7, 68, 72, 76, 79, 93, 106, 108
Stubbs, T.R. 6, 7, 8, 60, 61, 65, 66, 67, 69, 71, 81, 82, 86, 92, 95, 93
Switzer, Mr. 92

T

Taylor 14
Thunderstorm 53-54
Tobacco 48
Tobias 10, 35, 55, 56, 97
Trains 9, 32

A GOLD HUNTER'S EXPERIENCE

Turtles 24

U

U.S. Mail 27
Utah 103

W

Wagons, repair of 13

Wall Street 109
Watson, Joe 103, 104-105
Westport, MO 20
Whisky 28, 48
Wife, trading a pony
   for an Indian wife 50
Wolves 18, 19, 33, 34

COLOPHON

The Chalkley J. Hambleton, A GOLD HUNTER'S EXPERIENCE, *was printed in the workshop of Glen Adams, which is located in the quiet farming village of Fairfield, southern Spokane County in Washington state and one township removed from the Idaho line. This is an enlarged facsimile of the rare Chicago edition printed in a small run for private circulation. We added a fresh title page and this colophon. I regret that I was not able to find any information at all about the author, although I tried. The original copy was supplied by William Thorsen of Chicago, former owner of American Book Collector magazine. The added material was typeset by Pat Nigh and Dale La Tendresse with a Compugratic Editwriter 7300 using nine and eleven point Baskerville with two points of leading between the lines. Indexing was done by Dale La Tendresse. Photography/darkroom work was by Sylvia Fenich using a 20x24 inch Model 660-C DS camera and a 25 inch LogE film developing machine. The film was also stripped by Sylvia Fenich. Plates were made by David Hooper, who also printed the sheets using a 28-inch Model KORS Heidelberg press. Folding was by Garry Adams using a 22x28 three-stage Baum folding machine. Paper stock is 70 pound Island Offest, a Canadian sheet. Paper binding was by Glen Adams and Garry Adams. This was a fun project. We had no special difficulty with the work.*